PURE LUCK
FARMSTEAD GOAT CHEESE
CHÈVRE
PERMIT #: 489718
DRIPPING SPRINGS, TEXAS 78620
PURELUCKTEXAS.COM
CONTAINS: GRADE A PASTEURIZED GOAT MILK, SALT, CULTURE, MICROBIAL ENZYMES
LIRA ROSSA ARTISAN CHEESE
ASIAGO AMERICANO
INGREDIENTS:MILK, CULTURES, ENZYMES, SALT
KEEP REFRIGERATED
Lira Rossa Artisan Cheese, LLC @ Four E Dairy
784 Co Rd 251, Moulton, TX 77975
www.lirarossa.com
Lot No
Net Wt
Permit No 480320
Cheddar
OROBIANCO ITALIAN CREAMERY
SPICY SICILIAN
BUFALETTA
WATER
CHEESE
AF854569
GO TEXAN.
LIRA ROSSA artisan cheese
MOULTONZOLA
INGREDIENTS: Thermized Milk, Cultures, Enzymes, Salt
Lira Rossa Artisan Cheese, LLC @ Four E Dairy
784 Co Rd 251, Moulton, TX 77975
www.lirarossa.com
KEEP REFRIGERATED
Net Wt: lbs oz
Permit No: 480320
Lot No:
ARTISAN · ORGANIC · DOUBLE CREAM
ARMAGH CREAMERY
DUBLIN' DOWN
BRIE-STYLE
ESTATE · MADE IN DUBLIN, TX
CKC FARMS
17505 FM 32
Blanco, TX 78606
830-833-5669
Jalapeño & Artichoke
"Farm Fresh Goat Cheese"
Ingredients: Grade A Pasteurized Goats Milk, Culture, Organic Vegetable Rennet, Salt, Pickled Jalapeños, Marinated Artichoke Hearts.
HOME OWNED SINCE 1928 HOME OPERATED
Knolle DAIRY FARMS
FREE RANGE FARMSTEAD PRODUCTS FROM SANDIA,TX
Farmstead Fresh
Pasture Based
Grass Fed
Real Local
All - Jersey
100% WHOLE MI
"Our Cows live on
All Natural
Low Temp Vat Pasteuri
Knolle Dairy Farms I
973 CR 360 Sandia, TX
Plant code 480816
KEEP COLD, SHAK
One Quart
DOUBLE CREAM BLUE CHEESE
HOME OWNED SINCE 1928 HOME OPERATED
Knolle DAIRY FARMS
FREE RANGE FARMSTEAD PRODUCTS FROM SANDIA,TX
QUESO PANELA DOS VERDES
ING:WHOLE PASTEURIZED MILK.CULTURES.RENNET.SALT.CALCIUM CHLORIDE ROASTED SERRANO.GREEN ONION MADE BY KNOLLE DAIRY FARMS INC. 973 CR 360 SANDIA.TX 78383.PLANT CODE 480816.BEST BY 5/8/24. 10-12 OZ.
OROBIANCO ITALIAN CREAMERY
GO TEXAN
TUSCAN BLEND
BUFALETTA
FETA-STYLE WATER
BUFFALO MI
100% TEXAN WATER BUFFALO
RIVER WHEY CREAMERY
ANAL CHE
MOZZARELLA CHEESE
Ingredients: Pasteurized cow's milk, cultures, vegetable rennet (rhizomucor miehei), salt.
MOZZARELLA COMPANY
2944 Elm St., Dallas, TX 75226
214-741-4072
www.mozzco.com
Keep Refrigerated
GO TEXAN.
LIRA ROSSA artisan cheese
LATTERIA FRESCO
INGREDIENTS: Raw Milk, Cultures, Enzymes, Salt
Lira Rossa Artisan Cheese, LLC @ Four E Dairy
784 Co Rd 251, Moulton, TX 77975
www.lirarossa.com
KEEP REFRIGERATED
Net Wt: lbs oz
Permit No: 480320
Lot No:
CKC FARMS
17505 FM 32
Blanco, TX 78606
830-833-5669
Peach Chipotle
"Farm Fresh Goat Cheese"
Ingredients: Grade A Pasteurized Goats Milk, Culture, Rennet, Salt, Peach Preserves, Chipotle Powder.
OROBIANCO ITALIAN
GO TEXAN

# the Texas Cheese Trail

WITH

Otto's Cheese Shop

*photography by*

Marshall Tidrick

*The* BEST LITTLE CREAMERIES *in the* LONE STAR STATE

# *the Texas Cheese Trail*

*by* BOB & RACHEL PINEDA
*owners of* OTTO'S CHEESE SHOP

Bob and Rachel Pineda
ottoscheeseshop.com

Published in partnership
with *Texas Monthly* Studio

TEXAS
MONTHLY
STUDIO

Studio.TexasMonthly.com

Editors: Kendall Talbot, Kathy McFarland
Creative Director: Harrison Albrecht
Designer: Rachel Hudson

Photographer: Marshall Tidrick
Food Stylist: Ashleigh Amoroso

Writers: Amanda Albee, Amy Brinkerhoff, Sterry Butcher, Tina Danze, Marianne Dougherty, Leslie George, Tyler Hicks, Kate Hull, Karen O'Donnell Krajcer, Roger Munford, Andrew Paul, and Abigail Rosenthal

ISBN: 978-0-9724495-8-8

*This book is dedicated to all those cheese-loving,*
*hardworking, pioneering can-do people out there.*
*And to our four-legged ruminant friends,*
*for without them, there would be no cheese.*

▲
Pure Luck
Farm & Dairy
cheese spread

## Introduction



## Creameries

▲
Inside the cheese cave
at Veldhuizen Cheese

CHEDDAR 030724
REDNECK CHEDDAR
012924
CHEDDAR 042924

•TEXAS SAMPLER•
•LOVE BITES•
TASTE OF EUROPE
OTTO'S CHEESE SHOP
OTTO'S CHEESE SHOP
ROCO

# *It Began With a Dream...*

*written by*
Amy Brinkerhoff

What is it about Texas that inspires people to dream big? Is it the endless, rolling skies above the plains? The breathtaking vistas of the West? Perhaps it's something more biological—a spirit that infuses the blood, handed down from the explorers and ranchers of previous generations.

This notion of dreaming big isn't reserved for pioneers and settlers, and Texas-sized dreams often crop up in some of the most unsuspecting places. Take, for example, the dream hatched by Bob and Rachel Pineda of Otto's Cheese Shop. Nestled in the heart of the Texas Hill Country, this fantastic little fromagerie offers a variety of carefully curated, artisanal cheeses from around the world.

Vision for the shop grew from the Pinedas' love of all things cheese-related, from sampling new cheese varieties to learning the methodology and science behind the cheesemaking process. In 2021, the couple decided to open their artisanal shop and share that passion with their community.

As business grew and word spread, cheesemakers from across the Lone Star State contacted the shop. Bob and Rachel steadily increased the number of Texas cheeses offered at Otto's, and customers were amazed by the diversity of cheeses and the complexity of local flavors.

But when patrons began begging for more information about Texas cheeses, the Pinedas discovered a glaring hole in the culinary market—not a single, comprehensive resource on Texas cheeses and their makers existed. And just like that, the spark of a new dream ignited.

That flicker was further fanned after Rachel attended an American Cheese Society Conference and learned that Texas, despite being the fifth-largest dairy-producing state in the nation, has no official cheese guild or way of connecting cheesemakers and cheesemongers.

And so, these cheesemongers decided to set off on a truly Texan adventure, blazing a trail across the state visiting farms and creameries, mingling with animals, listening to the musings of makers, and learning about their unique cheesemaking processes.

It turns out that the Pinedas' passion for cheese is not theirs alone, for throughout the Lone Star State are farmers and makers who have channeled the Texas spirit to create some truly exceptional cheeses.

For Bob and Rachel, this book has been a labor of love and the realization of a dream long in the making. They are excited to share a truth that has been kept secret too long—the fact that some of the world's best cheeses and most talented makers can be found right here, deep in the heart of Texas.

*opposite*

Rachel and Bob Pineda, owners of Otto's Cheese Shop, New Braunfels, Texas

# Texas Cheese Trail Map

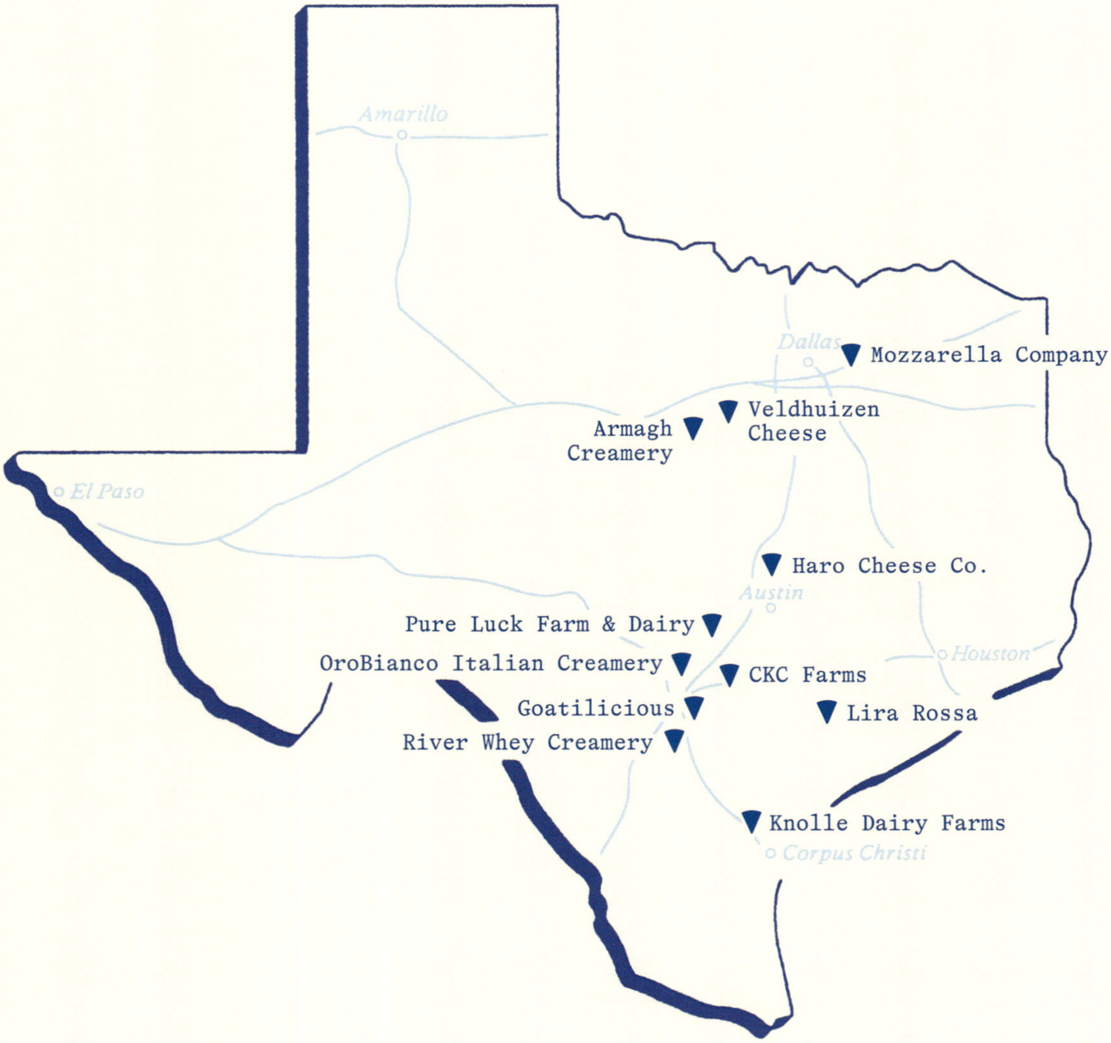

# *The Spirit of Texas Cheese*

*written by*
Amy Brinkerhoff

The history of cheese in Texas is long, but hardly storied. For more than a century, makers throughout the state have quietly produced cheese with little fanfare, honoring an age-old tradition relatively new to the Lone Star State when compared to other parts of the world.

Throughout most of the nineteenth century, commercial dairy products like milk and cheese were imported from northern states, though many living in small towns and rural areas used milk from their own dairy cows to make butter and cheese for the family.

As the number of immigrants increased in Texas so, too, did the variety of cheeses. German settlers introduced kochkäse, or "cooking cheese," to the state, a spreadable cheese made by cooking down days-old clabber, or curds and whey, with a bit of fresh milk. Pioneers from Mexico brought stretchy, milky queso asadero to the southwest regions of the state, kicking off Texas's deep and abiding love for queso in all its delicious variations.

The dawn of the twentieth century saw an increase in Texas dairy production, with a dozen creameries operating in the state by 1900 and 89 creameries by 1913. The industry has grown steadily since—Texas currently boasts more than 375 producer-dairy farms and enough cheese-from-scratch producers to rival Vermont.

Like Tex-Mex cuisine and Tejano music, today's Texas cheeses are decidedly unique and heavily influenced by their region of origin, a fact not surprising given the vastness of the state and the ten ecoregions within it. From Gulf marshes and Post Oak savannahs to the South Texas plains and the Piney Woods, the flora and fauna of each region have a hand in the flavor of its cheeses. This explains why the pungent flavor of a West Texas goat cheese tastes entirely different than one made in the heart of the Hill Country.

And it is precisely this uniqueness—this notion that, no matter where you roam in the Lone Star State, you'll never experience the same taste twice—that has grabbed the attention of Texas turophiles who revel in the variety of cheeses now readily available. Whether it's a traditional manchego-style wedge, a flavored chèvre, or a ball of water buffalo mozzarella, there's just no end to the cheese offerings found here, each one as unique as the state itself.

There's no question Texas cheesemaking has grown in popularity over the last century, but Texas cheesemakers are hesitant to call it an explosion or consider themselves trendsetters. That's just not Lone Star style. Today's Texas cheesemakers are a fascinating amalgamation—part alchemist, part artist, and part pioneer.

You'll be hard-pressed to find a maker who's looking to get rich from the process. What you will find is a fascinating group of passionate artisans who want nothing more than to do what they love on the land they love, and to create a product they can be proud of. And if that isn't the embodiment of the Texas spirit, we don't know what is.

# A Still Life

Texas is experiencing something of a cheese renaissance. Indeed, a surge of cheesemakers in the Lone Star State are creating award-winning artisan cheeses that rival those made in California, Vermont, France, and Italy. This is good news for cheese-loving Texans, who have a bounty of fresh, tasty offerings to choose from.

Indigo Ridge
*gouda-style*
River Whey Creamery
Caraway Cheddar
*cheddar*
Veldhuizen Cheese
Eldo
*manchego-style*
Haro Cheese Co.
Queso Quesadilla
*Mexican-style cheese*
Knolle Dairy Farms
Queso Oaxaca
*Mexican-style cheese*
Mozzarella Company

*owned by*
Ryan and Katie McDowell

*cheese*
Doublin' Down

# Armagh Creamery

## *Dublin, Texas*

With a dairy heritage in Northern Ireland that dates back nearly one hundred years, Armagh Creamery sets up its family operations in Dublin, Texas.

Dairy farming is in Ryan McDowell's blood. The co-owner of Armagh Creamery in Dublin, Texas, can trace his roots back to County Armagh in Northern Ireland, where his great-grandparents grew up dairy farming. After emigrating to the United States, they built a dairy in Petaluma, California, which the family owned and operated for ninety years. Ryan's father, Mike, began growing organic produce on the farm in 1995 and converted the dairy side of the operation to organic in 2005. When Mike and his wife, Linda, decided to move to Dublin, Texas, in 2010 to start an organic dairy farm there, they passed the baton to Ryan, who had grown up on the farm and was happy to take over the family legacy.

Then in 2019, Mike and Linda persuaded Ryan, his wife Katie, and their four children to relocate to Dublin to start a new family dairy and creamery. "We chose the name to honor the family legacy," Katie says.

◀ The McDowells have four children: Elly, Matthew, Ben, and Jack (pictured here)

Making cheese was a new endeavor for the McDowells, who had sold milk from their herd of dairy cows to co-ops but had never made any products of their own.

"Texas has a lot of wonderful hard cheeses, like cheddar and gouda, and there are great goat milk dairies making small batches of cheese," Ryan says, "but we noticed that there weren't many people doing soft-ripened cheese." Their brie-style cheese, Doublin' Down, fills a niche in that market. "We're pretty new at this point, but we hope to develop a reputation for quality artisanal cheese."

With construction of the creamery completed in December 2019, the couple was still tinkering with the recipe for Doublin' Down while making small batches of European-style yogurt. "The cream rises to the top," Katie explains of the yogurt's smooth, creamy texture. The idiom is apt for a small family business that has overcome obstacles to not only survive but thrive. For example, raising dairy cows in Texas has been challenging. "You've got a mild, temperate climate in the Bay Area," Katie says, "but in Texas where it's really hot in the summer and cold in the winter, animal management becomes more difficult."

In the springtime when the cows are grazing, their milk takes on a pale golden hue. The McDowells farm about five hundred acres of hay and silage free of herbicides and pesticides that they feed their herd in the winter when native grasses in Texas go dormant. To keep the microbiology of the soil consistent, Ryan plants rye and winter wheat in the offseason, a regenerative process designed to maintain a diverse plant

▼ *below*

Ryan McDowell practices regenerative farming to ensure something is always growing on the farm

► *opposite*

Armagh Creamery's herd of Jersey crosses grazing

base so the soil is always growing something. Theirs is also a no-till farm. Tilling the soil, he explains, releases carbon into the atmosphere, while he prefers to leave it in the soil where it can feed the root system.

"We have eighty cows that we milk daily," says Ryan, who describes them as Jersey crosses (mostly Jerseys, though some have been crossed with Holsteins). "Creamier milk is the gold standard for making the kind of soft cheese we do, and Jersey milk is known for its high butterfat."

That milk is less than 24 hours old when processed because it's pumped directly from the dairy to the creamery, which is about forty feet away. To make Doublin' Down, the milk is pasteurized and then inoculated with penicillium candidum, a mold that will ripen and flavor the cheese while forming a white, bloomy rind. "From cow to wheel, the whole process takes about thirty days," Ryan says with pride. "That's the cool thing about what we do. We're making our products fresh on the farm."

So, how did a Texas-based dairy farmer learn to make a brie-style cheese with a profile comparable to those found in Europe? The easy answer is trial and error, but it's also the right one. "That's part of the artisan cheese business," Ryan says, who met with a few cheesemakers, one of whom agreed to come down from New England to spend a few days at the farm. "He helped us get pretty close to what we were after, and we went from there, tweaking a bunch of recipes."

The thing about a good brie is that the flavor becomes more complex as it ages. A young cheese, which has a texture not unlike fudge, may be noticeably sweet, but a decidedly earthy and mushroomy finish begins to emerge as it ages and gets creamier from the inside of the rind to the center.

"It's interesting how cheese varies by region, capturing the flavors of the 'terra' where it's produced. They're more mindful of that in Europe, where there's a long history of cheesemaking, and we hope to bring that sensibility to the way we make cheese here in Texas," Katie says. "The brie you're buying here that's been imported from France is not the brie they're eating there. We're trying to give our customers the experience that European consumers have by creating something locally without preservatives and stabilizers." Ryan describes the process as "dynamic" precisely because the profile changes from beginning to end.

▲

*above*

Doublin' Down pairs well with jams, honey, nuts, and charcuterie

Ryan loves meeting customers in person at farmers markets and also offers tours of the farm so people can see what they do there. "I want our customers to see that our cheese comes from a herd of cows that are well taken care of by a family that takes pride in what they do."

"We're creating a legacy for our children," Katie says, "and I love that they've had a front-row seat, watching their parents and grandparents work hard and seeing where their food comes from."

At the moment, Doublin' Down is only available at a few local cheese shops and seasonally in larger retailers, but Katie is convinced that "we're on the verge of an exciting time in Texas where cheese is taking off" and that there's plenty of opportunity to expand into more restaurants and local shops. The couple has also talked about staying in the soft cheese world that they know but offering variations on the recipe they have perfected. A couple of options include putting more herbs in the rind to add a Southwestern flavor to the cheese or experimenting with dried cranberry in the fall.

Aside from the long history of dairy farming in Ryan's family that has been passed down from one generation to the next, what continues to inspire the couple is that they are building a pathway for farmers, something that's sustainable. "Having grown up in the Bay Area where we were inspired by cheesemakers we met there, we had access to high-quality artisanal cheeses," Katie says, "and that's what we want to bring to the people of Texas."

◀

*opposite clockwise from top left*

Cheesemaker Karla Martinez fills cheese molds

Ryan in his cheesemaking gear

About five days into the aging process, Doublin' Down starts to form its white, bloomy rind

*written by*
Marianne Dougherty

*owned by*
Omo Family

*cheese*
Baby Blue, Baby Caprino, Jalapeño & Artichoke Chèvre, Peach Chipotle Chèvre, Seasoned Feta

# CKC Farms

## *Blanco, Texas*

A family's love for their land, their animals, and one another are the secret ingredients behind these goat milk cheeses.

Chrissy Omo was fifteen years old when she learned the potential fate awaiting her pair of show goats, Ozzy and Cosmo. The bad news arrived during one of her high school Future Farmers of America meetings, and she immediately vowed to save her beloved animals from the chopping block. More than eighteen years later, it's safe to say the importance of caretaking has become a central aspect to not only her life, but to her entire family as well.

"Whether it's rain or shine, whether you feel like it or not, the dedication is in showing up for an animal that has no one else but you to help them," she says. "Having that responsibility of being a steward is a bittersweet challenge."

Now with two children of her own, Chrissy isn't quite as involved in the day-to-day activities at CKC Farms in Blanco, Texas, but she helped build the family's goat dairy into the beloved local staple it is now.

◀ Saanen dairy goat kids

The day-in, day-out work requires immense dedication from every family member—Kenny Omo oversees many of the ranch's operational aspects as director, while their brother, Conner, helps care for the more than 160 milking goats, the herd made up of Alpine, LaMancha, and Saanen breeds. Meanwhile, their mother, Adriana, handles much of the cheesemaking, and their father, Daniel, takes care of mechanical repairs and maintenance.

"The dynamic of working with family can be the most rewarding thing," Chrissy says. "You hear a lot of companies say they are 'family-oriented' or 'family-structured,' but we really are a family."

Way back when, however, it was simply a hobby for the Omos—Kenny recalls many failed attempts and cheesecake salvages during what Chrissy describes as their "mad scientist" era. As the number of successful, delicious creations increased, however, Adriana and the rest of the family wondered what they could accomplish together as a full-fledged business.

"It just kind of took off from there. It was a lot of fun, the huge family involvement and support," Kenny says.

Today, CKC Farms boasts a wide array of fresh chèvre: herb, jalapeño and artichoke, and a particularly popular peach-chipotle variety promising a sweet, smoky flavor. There's also an aged feta made with extra virgin olive oil, Mediterranean seasonings, and sun-dried tomatoes, as well as one of the undisputed family favorites—a lightly-aged, creamy baby caprino featuring an earthy, mushroom flavor similar to that of a camembert. Their sought-after Baby Blue is an aged cheese modeled after a gorgonzola-type recipe.

"They can eat a whole wheel by themselves," Chrissy says of her children.

Kenny and Conner even hint at a goat milk mozzarella they are tinkering with.

"It's an interesting process, but we have failed quite a bit with that one," Kenny says with a laugh.

Of course, the Omos toil for days before customers can taste any of these wonderful options. Fall and winter may sometimes afford the luxury of slightly later starts to the day, but mornings begin around 6 a.m. during the spring and summer months, when the mother goats give birth and begin raising the baby goats, or kids. Because the animals on CKC Farms are bred specifically for milking, they produce so much of it that they require near-constant supervision and maintenance. Conner is currently responsible for this task, which usually takes between an hour and an hour-and-a-half. Spring and summertime yield even more milk, so the regime greatly increases to ensure the "mama goats" (as Kenny calls them) stay healthy and happy.

While much of that fresh milk makes it over to the cheese process, quite a bit of it naturally must remain for—who else?—the kids, who are bottle-fed twice daily. What doesn't make it to the new generation is transferred to a bulk tank for cooldown and pasteurization. After pasteurization comes the culture phase, when vegetable rennet is introduced over another day before Kenny and Conner separate the curds and whey with cheesecloth. More intricate work may follow depending on the chèvre variety they are making at the time—and then comes the deliveries to clients in San Antonio, Austin, Houston, Fredericksburg, and elsewhere.

Alongside their steadfast caretaking regimes, the Omo family believes the Lone Star State's climate is what helps their CKC Farms goat cheese stand out from the offerings found anywhere else in the world: The Texas environment really makes its mark on the cheese produced at CKC Farms.

The CKC Farms diet for the goats is also markedly different from, say, goats roaming the Swiss Alps. In Blanco, while they are mostly grass-fed,

◄ *opposite*

Bottle feeding spring kids

▼ *below from left*

Daniel and Adriana Omo in their creamery

The "CKC" of the creamery name (from left): siblings Chrissy, Kenny, and Conner

BOBCATS

CKC FARMS
17505 FM 32
Blanco, TX 78606
830-833-5669
Seasoned Feta
"Farm Fresh Goat Cheese"
Ingredients: Grade A Pasteurized Goats Milk, Culture, Organic Vegetable Rennet, Salt, Extra Virgin Olive Oil, Sundried Tomatoes, Cracked Black Pepper, Rosemary, Bay Leaves, Seasonings.
CKC FARMS
17505 FM 32
Blanco, TX 78606
830-833-5669
Seasoned Feta
"Farm Fresh Goat Cheese"
Ingredients: Grade A Pasteurized Goats Milk, Culture, Organic Vegetable Rennet, Salt, Extra Virgin Olive Oil, Sundried Tomatoes, Cracked Black Pepper, Rosemary, Bay Leaves, Seasonings.
CKC FARMS
Seasoned Feta

the goats also receive a molasses protein, and they have access to a wide array of cedar, juniper, and oak tree bark for snacking. The pH balance of a goat's milk changes throughout the seasons—spring, for example, yields far more milk, but with less butter fat. In contrast, fall and winter offer less, albeit fattier, milk.

"It adds a lot of richness to the milk, the seasonality and the changes," Kenny says.

Apart from the rewards of raising their milk goats and crafting their delicious cheeses, the Omos say one of the biggest senses of satisfaction comes from the looks on customers' faces, be it at local farmers markets or elsewhere.

"We feel so much joy, because we actually make the product from start to finish," Kenny says. "Whenever we see someone buy a cheese, and their face is lit up, it gives us a sense of fulfillment. We know we're doing a good job."

Daniel Omo agrees. "People say, 'Well, I don't really care for goat cheese,' and then they try it and say, 'Wow! That's really good!'" he notes. "The proof is really in the tasting."

Borne out of nearly two decades' worth of hard work, cooperation, creativity, and family love, CKC Farms ranks as a top-tier Texas source for both traditional and modern takes on the rich world of goat cheese.

▲ *above*

Peach Chipotle Chèvre, which is soft and spreadable, is great with crunchy crackers and a crisp cider to match the acidity of the goat cheese

◀ *opposite clockwise from top left*

Conner pours the goat milk into a storage bulk tank

Chrissy uses her favorite cheese knife to cut feta curds

Jars of the CKC Farms seasoned fresh goat cheese feta

*written by*
Andrew Paul

*owned by*
Robert Ragels

*cheese*
Cajeta, Fig & Black Pepper Chèvre,
Jalapeño & Artichoke Chèvre, TommBoy

# Goatilicious

## *New Braunfels, Texas*

Goat milk healed their son, which drove this family to start their popular farmstead cheesemaking operation.

Robert and Claudia Ragels started making cheese twenty years ago as an afterthought. Today, their farmstead chèvre can be found in some of the best restaurants in San Antonio and Austin, and at regional farmers markets throughout the Texas Hill Country, where people line up for their creamy Goatilicious Fig & Black Pepper Chèvre and their one-of-a-kind tomme-style goat milk cheese called TommBoy.

"We started with goats to heal our son," Robert says. The eldest of their two boys, Robin, had severe eczema and chronic ear infections when he was five years old. "His doctor said he may be sensitive to a protein in cow milk and advised us to try goat milk, which contains a more digestible protein." At the time, the couple was living on two acres near Flagstaff, Arizona, with Robin and their baby, Ryan. They bought two goats and switched to drinking only goat milk. "He stopped getting ear infections, and his eczema cleared," Robert says. Then came the cheese.

◀ Goatilicious owner Robert Ragels at his farmstead creamery

"When you get goats for milk, there comes a time when your two goats turn into four goats, and suddenly you have an overabundance of milk," Robert says. "We had so much we were dumping it down gopher holes."

Robert started researching how to make cheeses, like mozzarella and chèvre. "I didn't have a culinary background, but I loved the chemistry of it. I'm still fascinated by how you can take a liquid and do a couple of things to it, and it turns into something that doesn't resemble what it started as."

In 2005, the family moved from Flagstaff to the rolling hills of New Braunfels, Texas, about thirty miles from San Antonio and an hour's drive from Austin. They bought fifteen acres of land rich in oak and cedar groves dotted by mesquite and some cacti. Beautiful, he says, but not necessarily great for raising goats. "The summers are too hot to grow enough food for a herd of goats," Robert says. "We thought we might retire from the goat business."

But he and Claudia happened to own the URL drinkgoatsmilk.com, from their business in Arizona, and whenever anyone searched the words "goats" and "milk" online, their phone number popped up. "We got all kinds of calls about goat milk. So we got back into it, starting with selling to friends and neighbors," Robert says.

Today, Goatilicious is still entirely family-run. The dairy has about 110 milking goats (150 total). Seventy percent of their business is in cheese sales. "I oversee cheese production, milk sales, and the day-to-day operation of the dairy," Robert says. "My son Robin works on the farm two days a week and sells at farmers markets three days a week, and Claudia does all the paperwork and handles all the animal sales and breeding. She knows the bloodline of every goat we have."

Understanding the animals that produce the milk is just one factor that sets a farmstead cheesemaker apart, Robert says. At his dairy, all the baby goats are bottle-fed from the day they are born, which allows the Ragels to control milk production and makes for a better-behaved adult goat.

"Almost every doe born here is going to be somebody's family milker on their farm, so they have to

▲ *above*

Goatilicious's popular cheese TommBoy at different stages in the aging process

be very people-friendly," Robert says. "A goat raised by a person is one hundred times more friendly."

Having worked with a range of breeds over the years, their herd today is made up of Nubians and LaManchas. "In my mind," Robert says, "Nubian and LaMancha goats have the mildest milk." Added benefits of those breeds: Milk from Nubian goats has the highest butterfat at 5 percent, and LaManchas are the easiest to work with, he says. But goat behavior also depends on how they are treated. "You have to break their natural built-in instinct of fear. The only way to do that is to be kind to them."

Feeding the goats what they like is part of that. To get around Texas's hot, arid summers, Robert imports alfalfa from Kansas to feed his herd, which they love. "In the Alps, the goats are moved from pasture to pasture throughout the year where they get clover and alfalfa and other grasses," he says. "We can't do that here." The high quality of the Kansas alfalfa, which has a 20–25 percent protein content, benefits the dairy as well. "The more protein a goat consumes, the more milk she produces," Robert says.

Farmstead goat cheese is also distinguishable in taste and texture because it's so fresh, made from milk produced on the premises. "Our milk goes from the goat to the tank room, where it's stored. Then, from the tank room to the cheese kitchen, which is no more than twenty feet away," he says. Some of the milk stays in the tank room longer, to be matured for the raw milk cheeses. Pasteurized milk moves on to the cheese kitchen much faster, where it's drained, flavored, and placed into molds to age.

◀ *opposite*

Baby Nubian goats smile for the camera

chael

Aside from its plain chèvre, with its mild, sweetly tart, buttery notes, Goatilicious offers a range of cheeses and products, including feta, the hard cheese TommBoy, and cajeta, its famous Mexican goat milk caramel, which customers love to put in coffee and on ice cream. And everything comes from the same milk—just one tweak in the process can lead to an entirely different product. "A temperature difference of five degrees at the end of the aging session or a time difference of an hour in the culture vat makes a big difference in the taste of the cheese," Robert says. "That's how delicate the cheesemaking process is."

"It just takes patience and accepting that you're going to make mistakes. I know animals, and I know milk, and nobody can make cheese the way I make cheese."

Robert Ragels | Goatilicious

Besides the best-selling Fig & Black Pepper, chèvre flavors include Garlic & Chives, Poblano & Garlic, and Tropical, containing toasted coconut and pineapple. There is always a range of seasonal pop-up flavors and products, like heart-shaped feta for Valentine's Day, Dill Chèvre in spring, Blueberry Chèvre in summer, and Pumpkin Cheesecake Chèvre in fall.

Robert is proud to say he taught himself cheesemaking by watching YouTube videos and reading books. "It just takes patience and accepting that you're going to make mistakes," he says. "I know animals, and I know milk, and nobody can make cheese the way I make cheese."

He points to the popular TommBoy hard cheese he invented, based on the recipes of tomme cheeses made in Switzerland and the French and Italian Alps. "I looked at different cheese styles from French and Italian dairies and added my take on it," he says. "We're the only place on the planet that makes TommBoy."

Robert shares that part of what drives him is that he cares about what his customers think. "The best part of my day is when I get a phone call or a text from a customer out of the blue, telling me they love the cheese or the milk," he says. "That's why we're here, in the end. Goat milk healed our son: We believe in the benefits of goat milk. We're doing this for our customers."

◄ *opposite clockwise from top*

Robert's goats line up in the milking room

Farm dog Selma, an Akbash mix, keeps a watchful eye on the herd

TommBoy, a hard tomme-style cheese, is cut into wedges for sale

*written by* Leslie George

*owned by*
Joel Haro

*cheese*
BTX! Jack, Camembert, Eldo, Red Handed

# Haro Cheese Co.

## *Bertram, Texas*

Classically-trained chef Joel Haro focuses on flavor with his array of raw milk cheeses.

The word "chef" awakens sensory images of a busy kitchen: the clatter of pans, the smell of citrus and spices, heat from the grill, shouts of orders, and the focused attention of white-jacketed culinarians spooning sauces over plates. But in the Central Texas town of Bertram, classically trained chef Joel Haro stands alone in a walk-in cooler, methodically flipping four-pound rounds of raw milk cheese over pine wood planks.

While the sole owner and cheesemaker of Haro Cheese Co. enjoyed some exposure to cheesemaking at the Culinary Institute of America (CIA) in New York, it was at the multitude of local, upstate creameries where Joel discovered an appreciation for artisanal cheesemaking. "I started to realize there's a whole world I hadn't seen," he says. "The most important thing the CIA pushes on you is taste. Taste everything. That's how you get an educated palate. That's really what sparks the ideas."

Years later, back in Austin, Joel embraced a new idea. A Waco-area creamery was for sale, the perfect opportunity to marry his passion for process with a beloved product.

◄ Haro Cheese Co. owner Joel Haro

Apprenticing under the retiring cheesemakers, Joel learned how to create three different kinds of cheese—a manchego-style, an aged jack, and a majorero-style—all using raw cow milk, milk that has not been pasteurized.

"It was fun to learn how to influence a product by applying different variables: temperature, pressure, and time. I liked the process, a physical component, a cerebral component—all of it was fun, and you got cheese in the end, which was a bonus," he says.

Eventually, Joel purchased a new cheesemaking facility in Bertram, and after the buildout was completed, finally began making cheese in the summer of 2022.

The immediate challenge? After years of preparation, Joel was eager to get his cheese to the public, but because of FDA policies, the raw milk that defines his cheeses' characteristic flavors would have to age for sixty days. His solution? Create a delicious, small batch cheese that could hit the shelves in the meantime: a camembert—Joel's favorite from his time in New York—made with pasteurized milk and cream. He and his wife "put one away every night for a week," he says. "It was wonderful." Yet, as he predicted, "once the raw milk was ready, that cheese was what the stores wanted."

Despite the extra time one must take with raw milk cheeses, Joel believes the final products are well worth the effort. "The biggest advantage is flavor. With raw milk you get the benefit of the full complement of microbes—the bacteria, the yeast, and the mold—and that is going to really influence your flavor profile of the cheese. It adds complexity to it." While pasteurization kills harmful bacteria, it also destroys the good, which would otherwise play a beneficial function in gut health. Joel understands the appeal of predictability that pasteurization affords, but ever the culinary scientist, he prefers the variability of flavor one can create with a live ingredient.

Once a week, Mill-King Market and Creamery delivers fresh, raw milk to Joel's facility in Bertram. The third-generation, farmstead dairy processor in McGregor has been caring for pasture-raised, grass-fed cows since 1941. In addition to being unpasteurized, the milk is non-homogenized, meaning that it hasn't been

▲
*above*

Haro's BTX! Jack, which placed third in "American Originals" at the 2024 American Cheese Society Competition

pressurized to break up and disperse the milk fat globules which would otherwise cause milk to separate. "You get this nice layer of cream," Joel explains, holding out his hands to suggest layers, "and you get to see what real milk is."

The day after the milk arrives, the cheesemaking begins. After warming up the milk, he introduces a culture, which drives the flavor of the cheese. Once in contact with the milk, the freeze-dried bacteria wake up and begin consuming the lactose (milk sugars). Next, Joel adds rennet, a coagulating enzyme that firms up the milk. When the milk has reached desired firmness, he cuts the gelatinous curd, allowing the liquid whey to separate. After draining the whey, he places the curd in a mold, where the last of the whey is expelled under pressure. Joel follows a specific process and pressing time for each of his three different cheeses before removing them from the molds, placing them in a 24-hour brine, and transferring them to the "aging cave."

No, not a literal cave, Joel says. "It's difficult to create a cave environment underground in Central Texas because of the limestone." Instead, the cheese wheels are aged on pine wood planks inside of walk-in coolers and flipped every two to three days. "You lose some of the terroir by having that artificial cave," he laments. Depending on the type, the cheese will sit for another three to six months until it matures, making Haro one of the few creameries in Texas to age cheese for such a long time. The longer the aging time, the drier the texture and more concentrated the flavor.

◀
*opposite from left*

Rennet is added after the milk has ripened with the culture

Once the milk sets, the curd is cut

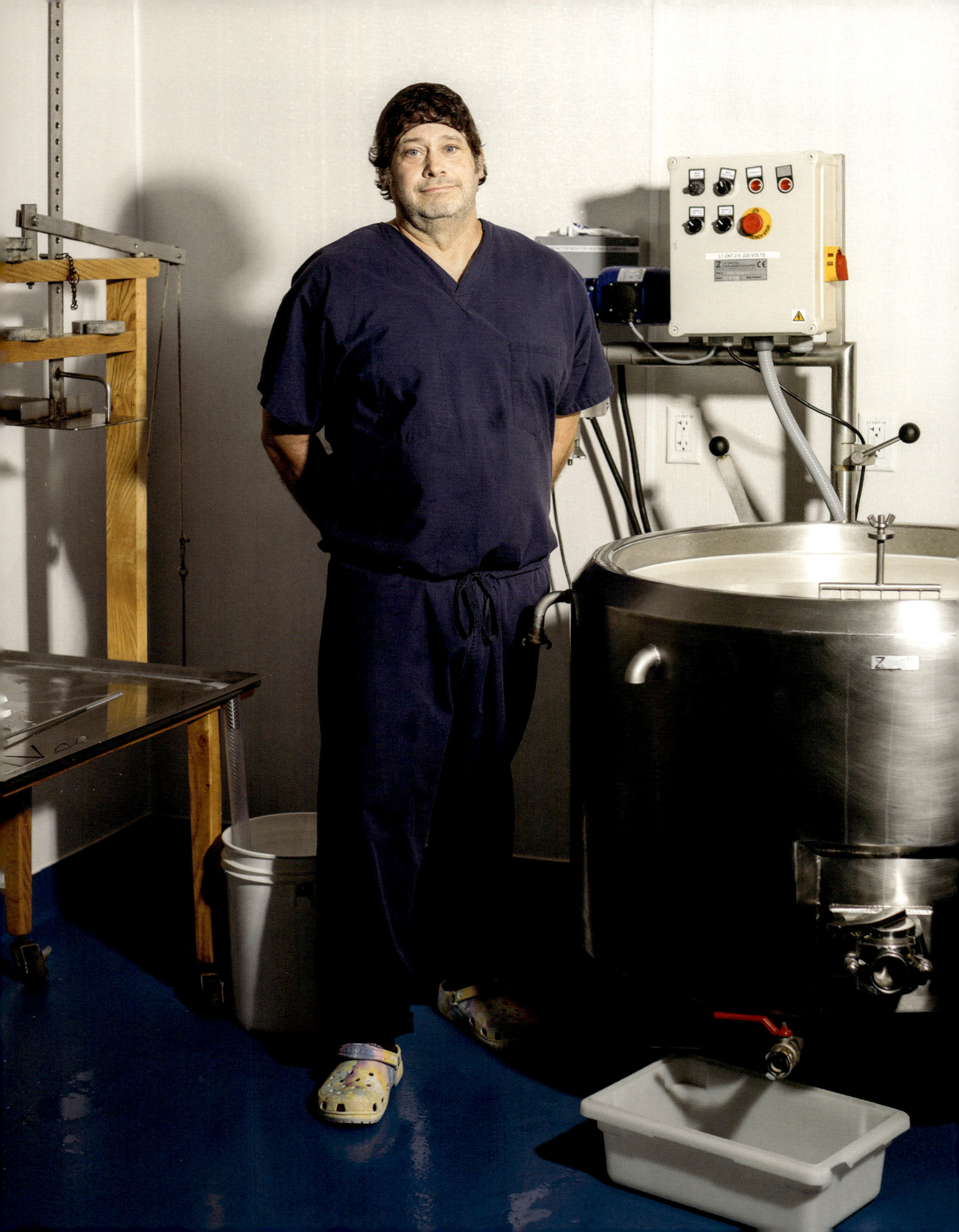
L1 CKT 2,4 220 VOLTS

Joel completes the process of cheesemaking, save the aging time, in a little over one day. After thirty days, the natural rind is rubbed in olive oil and another ingredient that imparts color. Like a traditional manchego, Haro's Eldo displays a grass belt with a black ash finish. The jack, pressed in a Kadova mold, is rubbed in cocoa, giving the cheese a smooth, brown finish that looks like stone. Haro's newest cheese, Red Handed, is a majorero-style cheese with a traditional pressed display of quilted diamonds finished in bright red paprika.

"I'm a big proponent of sourcing local. I want to involve things from the area, whether that's local beers and spirits or honey and meats."

Joel Haro | Haro Cheese Co.

If pressed to choose a favorite, Joel would pick the Eldo, which won first place in the Hispanic- and Portuguese-style cheese category at the 2024 American Cheese Society Competition. "It's been the most refined, and I'm really proud of where it is now," he says. The cow milk gives the cheese a creamier flavor than traditional sheep milk, and he doesn't know of anyone else making a manchego-style cheese with raw cow milk. When serving and selecting accompaniments, Joel looks for a contrast in texture as opposed to flavor. "We just eat it by itself, and I always prefer more plain crackers." For a festive presentation, Eldo "does great on a charcuterie board," and its creaminess pairs well with a good local honey.

"I'm a big proponent of sourcing local," Joel says. "I want to involve things from the area, whether that's local beers and spirits or honey and meats." Hand-in-hand with sourcing local is supporting local. Whey, a natural waste product of the cheesemaking process, is "pure protein" that can be used to supplement livestock feed. "I actively have multiple area farmers come by and pick up buckets of whey. It's a great way to upcycle, to close the loop on the sustainability end of it."

As it was in culinary school, collaboration remains essential for learning and growth. "That's what gets me going," Joel explains, "collaborating, learning, going somewhere new and eating as much as you can in that area—just seeing how people are doing things and then using what you have to make it your own." Looking ahead, Joel would like to expand his education with another tour of Vermont's artisanal creameries. It was there, shortly after he began producing his own cheese, that he recognized the stark contrast between small batch cheesemaking and the highly automated processes of large-scale, commercial creameries.

"The more I learned about cheesemaking, the more I appreciated the history and tradition," he says. "I like the idea of preserving the craft."

◀ *opposite*

Joel stands between his vat (right) and cheese press, both made and imported from the Netherlands

*written by*
Karen O'Donnell Krajcer

*owned by*
Joe and Christina Knolle

*cheese*
Chihuahua, Manchego, Panela, Quesadilla

# Knolle Dairy Farms

## *Sandia, Texas*

Approaching its one hundredth anniversary, Knolle Dairy Farms is the passion project of Joe and Christina Knolle, whose creamery is the lifeblood of their historical Corpus-area farm.

At the peak of Knolle Dairy Farms' production in the 1960s, brothers Henry and Ed Knolle never imagined that their 12,000-acre dairy, powered by what their milk cartons claimed as "the world's largest herd of Jersey cows," would be funneled into a small operation that thrives on artisanal Mexican cheeses. During that era, when more than one hundred other dairies existed in South Texas, the Sandia-based farm employed and housed 250 families who helped with managing a herd of eight thousand Jerseys, a breed known for producing nutrient-dense milk with a high amount of butterfat.

Five decades later, Knolle Dairy Farms is one of the only remaining dairies in South Texas. The reason for the diminishing number of dairies is a common story. As grocers started moving to nationally distributed brands, the family "saw the writing on the wall," as present day owner Joseph "Joe" Knolle Jr. says. His grandfather Henry and great-uncle Ed sold the milk plant to Hygeia Dairy in 1972, the beginning of a series of sales that eventually funneled into now defunct Dean Foods. In the process, the Knolle family name evaporated from legal paperwork and grocery shelves, just like other regional brands. As a result, over the next two decades, many of founder Almond Knolle's heirs divested from the dairy business established in 1928 to pursue other endeavors.

◄ Knolle's panela is perfect for grilling, eaten on corn tortillas with fresh cilantro and jalapeños

Joe is direct when he says "most family farms are becoming a thing for the history books." He and his wife bought the last of the farm's original herd when his dad, Joseph Knolle, announced his retirement at their annual Easter picnic on the Nueces River in 2012. Six months prior, Joe had married Christina, an international flight attendant who decided to "just go with it" when Joe wanted to keep the four-generation-spanning family business going.

They agree it was the end of their honeymoon. According to Joe, they "never worked so hard in their lives losing so much money," and he jokes that he sometimes wondered if Christina was going to run him over with her tractor. Even though they were participating in the federal milk order and continuing to incrementally downsize, they saw black ink only once over the next ten years. "Every other year was a horrific loss," Joe says.

Finding a pivot that Christina could get on board with was essential, even more so after they agreed to put a deadline on the experiment. As it happens, Christina had always enjoyed eating cheese and already had the prerequisites of a skilled cheesemaker: a love for cows, the curiosity to learn, and an interest in the simple beauty of Jersey milk.

When she started experimenting with a home milk pasteurizer (a Christmas gift from her father-in-law), Christina's first few batches of rubbery cheese were less than desirable, but she kept playing around, studying, and a taking advantage of a particular perk of her job: the ability to visit grocery stores in the Netherlands, Brazil, and Argentina to sample cheeses during layovers. By late 2020, the Knolles got a permit to produce and sell cheese. Word spread to the community, who were already driving to the farm for "agritainment," as well as exceptionally thick milk, yogurt, beef, chorizo, and other products available for purchase in a kiosk operated by the honor system.

Within a year, the Knolles were met with another bump in the road to becoming financially sustainable when inflation began affecting the price of commodities, or cattle feed. The next pivot was a blessing in

◀ *opposite*

Joe (center) and Christina Knolle with Uncle Pearson Knolle

▲ *above*

Cows munch on coastal dry chop, alfalfa, and sorghum silage

disguise as they moved to a vertically integrated, grass-fed farm, with the herd feeding off of coastal bermudagrass, clover, and their own curds. It was another move that no one thought would work, and it took about eight weeks to wean the cows off commodities designed to increase milk production. The process was "like bringing them down off a drug," Joe says, and resulted in 60 to 75 percent less milk. But the new grass-fed milk had more concentrated components, like keratin, which lent to even creamier, richer, almost buttery cheese.

Also in 2022, Christina took a cheesemaking class at Penn State University, and she was additionally receiving guidance from mentors like Hector Sanchez, a well-known cheesemaker from the Rio Grande Valley, and Dr. Ronald Richter, a retired food science professor at Texas A&M University. Operations eventually became more than Christina says she could handle with a full-time job, so she trained a neighbor, Dale Golden, now head cheesemaker, who "really poured his heart into it" and turned it into an art form.

Knolle's four main cheeses include chihuahua, manchego, panela, and quesadilla. Within those varieties are infusions guided by the couple's appetites, like panela with pasilla chile, quesadilla with chipotle pepper, and the Wholly Mole' It's Queso Fresco, made with mole sauce. The bestseller from the beginning has been the fuerte quesadilla, with roasted piquín peppers, garlic, and green onion. Joe is half-Mexican and loves this tasty product of his heritage, but more than that, he wanted to create cheeses the community would like. They're all now made by Dale

Knolle

▲

*above*

Knolle's Queso Fresco Fuerte is made with chile piquín, green onion, and roasted garlic

in a processing facility that Joe and a few employees constructed with equipment rummaged together from a two-year scavenger hunt.

Beyond the full-fat, grass-fed Jersey milk, other factors make Knolle's Mexican cheeses exceptional, like the fact no milk powders and very little salt are used, and the cheeses are vat pasteurized at a low temperature, which prevents denaturing of the milk. The sourcing of their cultures, or the bacteria that coagulates cheese, is another component. Christina describes the creation and distribution of cultures as an art form, like the cheesemaking. She points to the distribution of global cultures as a means to putting Texas cheeses on a playing field that can compete with anywhere else in the world.

◀

*opposite clockwise from top left*

Stirring the curds

Cutting the curd with a special curd knife

Dale prepares to clean lines after milk has been transferred from pasteurizer to cheese vat

Knolle's fresh cheeses come with a short shelf life, one reason the Knolles prefer to stick with local distribution around Corpus Christi, and in the future, possibly San Antonio and Victoria. They prefer the local community focus—unlike most businesses, growth isn't a priority.

"We have a social responsibility to try and provide the best wholesome, natural dairy products that we can for our community," Joe says. "Why should I be looking anywhere else when I've got to take care of the folks in my backyard?"

The best part for Christina—whose animals have pet names, like Clover and Boo Boo, and who has been known to place ailing calves in their home bathtub for rehabilitation—is the cows. As she says, "The whole reason we started making cheese is because we love animals."

*written by*
Amanda Albee

*owned by*
Andrea Cudin

*cheese*
Caciotta, C'mon Berto!, Crescenza,
Latteria, Moultonzola, Ricotta

# Lira Rossa

## *Moulton, Texas*

Italy-born Andrea Cudin blends the traditions of cheesemaking borrowed from his home country with American ingenuity to inspire flavorful cheeses.

◀
Pecan smoked ricotta and varieties of caciotta: plain, black pepper, red pepper, oregano, and truffle

Andrea Cudin was a man homesick for cheese.

Andrea is a native of the Friuli area of Italy, a pocket of about one million people snugged against Austria and Slovenia to the north and east and the Adriatic Sea to the south. Think snowy Dolomite peaks, red-tiled roofs and centuries-old villages, seaside castles, terraced vineyards cascading down verdant hillsides. And cheese. Really good cheese. Sweet, milky ricotta. Soft, creamy caciotta. Aged, aromatic latteria.

"I come from a region of Italy with the highest consumption per capita of cheese," Andrea says. "When I moved to the U.S., I realized I was missing cheese."

Cheese had not previously been a preoccupation for him. He did not come from generations of venerated cheesemakers. Andrea, in fact, had earned a PhD in philosophy and had a thriving career leading the office of international student affairs at an Italian university.

Despite his cool job that took him all over Europe, and his happy marriage to his wife, Jillian Scheumack Cudin, a Texas gal from Victoria, Andrea was restless. Their first baby was on the way. "I wanted to see something new in my life," he recalls. Four days before his fortieth birthday, in 2012, the Cudins moved to Victoria.

"I was really looking forward to something new," he says of their big shift. "But I expected that whatever new that was coming was still in my field." Academic international affairs jobs based in Victoria proved pretty scarce, though. And while his father-in-law hired him to work in the family contracting business, he discovered the construction trade wasn't his natural calling. He had a baby, a wife, a mortgage. Good thing, Andrea says, "in this country, there are second chances."

The first batch of latteria, a traditional Friulian cheese, came out of the Cudins' kitchen. One experiment begat another, with a five-gallon pot dedicated to cheesemaking and a spare fridge in the garage that was repurposed into an ersatz aging room. Andrea watched YouTube videos and read voraciously about technique, fermentation, pasteurization, starters, curds, and rennet. He made mozzarella next, and then ricotta, which is a by-product of latteria. Over the course of about two years, what started as a once-a-month hobby to reconnect him to flavors from his homeland evolved into a more serious endeavor that included stints here and there working with mentors at creameries back in Friuli. Andrea's efforts to make alpine-style cheeses in the brushy flats of South Texas were stymied by the quality difference in the milk that was available to him. After all, excellent cheese comes from excellent milk. He was hard pressed to equal the milk from Italian cows that grazed upon the mountain grasses and wildflowers of Friuli.

But then came the Chaloupka family. A third-generation dairyman, Chad Chaloupka's family farm is located in Moulton, sixty miles from Victoria. Faced with the possibility of losing their dairy, the Chaloupkas boldly switched to Jersey cows and intentionally diversified their offerings. Instead of selling all their milk for generic commercial use, they also began selling raw

▲

*above*

Chad Chaloupka and his family have run Four E Dairy since 1953

milk directly to a small but avid market of individual consumers. "With commercial milk, no one knows who made it," Chad explains. "With the raw milk, this is the first time we presented something that actually had our name on it."

Word of mouth led Andrea to buy the Chaloupkas' raw milk, and very quickly a partnership bloomed. Those gentle, doe-eyed Jerseys grazing in Moulton's pastures, so content and well-tended by the Chaloupka family, produced high-fat milk of outstanding quality. "Incredible milk," Andrea says. After a simple handshake deal, the dairy added onto the milking barn and created a creamery for the Italian. Chad Chaloupka didn't know the first thing about alpine cheese, but he keenly understands the pride in things well done. "Andrea's name is on the cheese, and we promote a product made from our family farm," he says. "It's two families with the same idea coming together."

The leap of faith from both parties still astounds Andrea. "In small town Texas," he marvels, "you can find people who are willing to trust just from having a good impression of someone."

Andrea dubbed the creamery Lira Rossa, or red lyre, as a nod to a book of Italian poetry owned by his Uncle Berto. Since 2016, Lira Rossa has produced its mainstays of latteria, mozzarella that's made daily, and delicate ricotta fresca, which Andrea sometimes smokes with pecan wood for a bit of Texas flair. Among the other offerings is the semi-soft caciotta, with rosemary or red pepper or seasonal variations, or the mildly tangy stracchino that's eaten alone or accompanied with a little honey.

◄

*opposite*

Cheesemakers Emil Martinez (left) and Andrea Cudin

▲
*above*

Latteria, a mild and buttery Italian-style cheese, pairs well with the juicy taste of olives and pickled carrots

The Formadi Frant is a Friulian choice of cheeses that are blended with juniper berries or horseradish. Texans particularly favor C'mon Berto, a camembert cheese with a brie-like texture named, once again, for his uncle.

"There are layers of flavors with Lira Rossa that you don't normally find in cheeses Americans have access to," Andrea says. "Here, there is a depth of flavor. Grass through the cow becomes milk—nothing's getting in between, no preservative, just a natural evolution from grass to a piece of cheese. The full spectrum of flavor is so much wider, reflecting where it comes from—the natural environment where it originated."

His ardent customers find Lira Rossa products at the little farm stand in Moulton, at farmers markets in Austin and Houston, and at a few specialty shops and restaurants in Texas and elsewhere. It's a small operation; Andrea works with two helpers while Jillian handles the books and taxes. He's had offers to go bigger, expand distribution, and widen availability. That's tempting, but Andrea says no thanks.

"That's not what we want to be. What we're trying to do is bring back good food and good food traditions. We're trying to create a sustainable environment for feeding people in a healthy, sustainable way," Andrea says. "You should be eating your local products, so the farmer next door can thrive. If your neighbor thrives, everyone can thrive while eating well. It's not about making the better cheese—that's just a consequence of a good, healthy way to do things."

◀
*opposite clockwise from top left*

Making caciotta: Andrea first cuts the curd

Filling molds with curd

Beginning the aging process

Flipping the curd from the mold

*written by*
Sterry Butcher

MOZZARELLA
COMPANY
Cheese made the
old-fashioned way - by hand.

*owned by*
Mauricio Travesi

*cheese*
Bierveza, Caciotta Ancho Chile, Dolce Habañero, Hoja Santa, Menonina, Mozzarella

# Mozzarella Company

## *Dallas, Texas*

Working in the tradition of Italian cheesemakers, the small team behind this Dallas-based creamery have been delivering celebrated cheeses for more than forty years.

In a true credit to his Italian heritage, Mauricio Travesi is a Renaissance man. In a life spanning nearly seven decades, the endlessly friendly Mauricio has been a creative director for commercials and other ads, a painter, and an airline pilot. But his favorite occupation, the job that'll likely be his legacy, is decidedly more tasty.

Mauricio is the owner and president of Mozzarella Company, a Dallas fixture that has won more than 150 awards and worked its way into the hearts of cheese lovers across Texas and the country. The shop was opened by legendary cheesemaker Paula Lambert in 1982, and Mauricio has been there for most of the years since.

"When I was working here in the 1990s, we would get letters from people who wanted to buy the factory," Mauricio says. "Paula kept them in a folder, and I asked her if she would ever sell. She wasn't sure, but I was sure of one thing: I knew I wanted to be here when she sold it. And I was."

◄ Mozzarella Company owner Mauricio Travesi

Mozzarella Company is ensconced in a street corner in Deep Ellum, a Dallas neighborhood best known as a music hub.

In a way, the location makes perfect sense: The vibrant flavors cooked up by Mauricio's team are a perfect complement to the cultural mishmash of murals, music, drinks, and food just outside the shop's doors.

When you enter, you feel like nothing has changed since 1982; the winding building has the kind of homey, accessible feel only few places can authentically capture. But authenticity is their thing.

Using just a pair of old, European-style cheese pasteurizers, the lean Mozzarella Company crew ships an average of fifty to sixty boxes of cheeses across the country every day. In addition to Mauricio, there's Vice President and owner/partner Ross Adami, a driver, and seven cheesemakers.

Mauricio bought the store from Paula in 2018, and he relishes wearing multiple hats on a daily basis. Sometimes he's a driver, and on every other Saturday he's a tour guide, providing joke-laced tours of the shop's cozy trappings, followed by a cheesemaking class.

"I spend a lot of time focused on shipping and production, so it's a real treat to be in front of the customers, telling them about our history and seeing their faces light up when they try some of our cheeses," he says, a mischievous grin flashing across his face. "I want to find out which one

▲

Mauricio with (from left) Elena Mercado, Leticia Villanueva, Carmen Lopez, Blanca Torres, and Judith Bailon

▲
*above from left*

Tossing fresh scamorza into the brine

Scamorza that has been smoked over pecan shells

is their favorite, so I can see if it's the same as mine." (For the record, it's the Queso Menonina, which he swears is even better than the mozzarella.)

Mauricio has swept-back silver hair and eyes that dance with laughter throughout those weekend tours. A masterful storyteller, he speaks softly and has a tale for every nook and cranny of the building and, of course, the cheeses too. Take, for instance, the hoja santa leaves his team uses to wrap the cheese of the same name. These Mexican pepper leaves are prevalent in Texas, so Mozzarella Company invites customers to bring in bags of hoja santa from their yards in exchange for a free wheel of cheese.

"In eight years, they'll own the place," Mauricio quips, referring to the many customers who eagerly bring in leaves to claim their prize.

In between the jokes, he tells the story of the Italian artisans whose photos hang from the walls above the pasteurizers. These are the figures who taught Paula Lambert how to make cheese when she spent five years in Italy studying the language and art history. She, in turn, taught Mauricio and her team.

"When I came to the U.S., I started at the bottom," says Mauricio, who was introduced to Paula by friends after leaving his pilot job in Mexico City. "I thought about leaving every few years, and once I did, but I always felt called back."

He pauses.

"It was probably the cheese. Or because Paula kept calling me."

Having such a small team gives the Mozzarella Company crew plenty of time to experiment. One year, a cheesemaker accidentally added an

HAND WASH
ONLY

unnecessarily spicy ingredient to a mix of cheese in the middle of the pasteurization process. Rather than toss the whole batch before a big Christmas delivery, Paula leapt into action, adding various ingredients until she found a taste she could approve of. Mind you, the cheesemaking process is lightning-fast with little to no room for error, so she had to work quickly to concoct something acceptable.

The delivery went out as planned, and when the shop didn't hear from any upset customers, they figured a crisis was averted. Roughly a year later, the phones were ringing even more so than usual in the run-up to Christmas.

"Nobody knew the name of it," Mauricio recalls, "but they all wanted the 'Christmas cheese.'"

Now the Christmas Cheese has been added to the store's selection all year long, and its mix of sharp flavors and subtle spice is anything but out of place.

## "It's a real treat to be in front of the customers, telling them about our history and seeing their faces light up when they try some of our cheeses."

Mauricio Travesi | Mozzarella Company

Stories like this are indicative of both the creativity and the hard work inherent to Mozzarella Company and Texas cheesemakers as a whole. Sure, Paula and the team she built are maestros in the mold of the Italians who came before them, but their establishment wouldn't be what it is today without years of her elbow grease.

"She used to putter around Dallas in a Pacer, selling cheese to any store that would have her," Mauricio says. He smiles fullest when relaying the story of Paula confronting the chef at the Mansion on Turtle Creek—an iconic upscale hotel in Dallas—and telling him, "This is the cheese you need to be using." He tried it, then agreed with her.

Now the creations she spearheaded are on sale in stores and markets throughout Texas, Arkansas, and Louisiana, and even though she has retired, she still occasionally visits to offer advice.

Not that Mauricio needs a whole lot of that.

The Renaissance man has mastered yet another profession, and he shows no signs of slowing down.

"Sales are good, but stable, and it's been like that for years," he says, noting his desire to one day have Mozzarella Company cheese on sale in national grocery stores.

"I just want to bring more cheese to more people."

*written by*
Tyler Hicks

◀ *opposite clockwise from top*

Making queso oaxaca, a string cheese

Nuts and dried apricots balance the earthy notes of the mild Deep Ellum Blue

Mozzarella is hand-formed into balls

*owned by*
Phil Giglio

*cheese*
Bufaletta, Buffalo Mozzarella, Blanco Fresco,
Stracciatella, Yellow Rose

# OroBianco Italian Creamery

## *Blanco, Texas*

Producing tender, creamy mozzarella with rich water buffalo milk proved challenging, but the team behind this Hill Country creamery has cracked the code.

Housed in an old dry cleaner building in the tiny town of Blanco, OroBianco is the first and only creamery in Texas to make cheese from water buffalo milk, a richer cousin to cow milk. It's also one of the only dairy-creameries in the country making water buffalo mozzarella, a Southern Italian delicacy that's notoriously difficult to produce; the challenge has eluded—even bankrupted—more than a few start-ups.

OroBianco's creamery, dairy, and shops are the brainchild of an unlikely newcomer to cheesemaking: Phil Giglio, an attorney who longed for an agrarian lifestyle while practicing securities law at a Chicago firm.

Although raised in rural New York, Phil had no agricultural experience. But when his law firm allowed him to work remotely, he vowed to trade his high-rise condo in Chicago for a ranch far away and pondered what he'd raise. Phil recalled his obsession with mozzarella di bufala during the months he lived on Italy's Amalfi Coast while studying for state bar exams; he had toured twenty creameries and dairies in Campania—where water buffalo abound—for the sheer pleasure of tasting the creamy queen of mozzarellas. That memory, coupled with his family's roots in the region, inspired Phil: He would raise water buffalo in Texas, where the warm climate suits the majestic, bovine giants, and he'd make buffalo mozzarella to sell to Texas Hill Country wineries.

◀ OroBianco's buffalo milk mozzarella served with fresh peaches, heirloom tomatoes, and basil, drizzled with olive oil and balsamic vinegar

"I figured 'no one's doing this, so I can't possibly be the worst at it.' Looking back, it's crazy, but it worked," Phil says. He christened his business OroBianco, which means "white gold"—the nickname for water buffalo milk in Italy. (He liked the similarity with "black gold," slang for another Texas product: oil.)

Phil bought a ranch near Fredericksburg and located a herd of four hundred pasture-raised water buffalo, reared by Jason Peeler of Peeler Farms, in Floresville. The herd wasn't for sale, but Jason agreed to partner with Phil on OroBianco. In 2018, they built the dairy in Floresville, which Jason manages, and began importing Italian Mediterranean water buffalo semen for breeding. To transport the fresh buffalo milk from the dairy to the creamery, they bought a milk truck.

Meanwhile, Phil retrofitted the dry cleaner building he'd bought in Blanco to make a creamery. His mother, Vicki Devendorf, an accountant, moved to the Hill Country to handle OroBianco's finances.

Making buffalo mozzarella proved a bigger challenge than Phil had imagined. A ten-day, hands-on class in Italy stoked his confidence. But for six months afterward, Phil and his skeletal crew failed to make a single successful batch of mozzarella at OroBianco. "The Italians told us, 'You will never make buffalo mozzarella in under five years.' It's easy to make a subpar mozzarella; it's really hard to make good mozzarella," Phil says.

▶ *opposite*

The water buffalo herd resides in Floresville

*photo by Jody Horton*

▲

(from left) Tammy Lopez, Adam S. Thompson, and Lori Thompson

With nearly twice the fat content of cow milk, buffalo milk can yield a more intensely flavored, richer mozzarella. It's also a more challenging base for mozzarella-making. The alchemy of turning such a rich, high-protein milk into a soft-textured, stretched cheese is as much an art as it is science.

Phil refused to give up on the mozzarella. But with close to $2 million invested in OroBianco, he needed a product to sell, fast. To keep the business afloat, he pivoted to making small-batch gelato, selling it alongside espresso drinks at a storefront he built inside the creamery. The gelato was an instant hit and remains a core part of OroBianco's business.

Meanwhile, Phil added another partner—Adam Thompson, a cheesemaker from the Rio Grande Valley with more than ten years of experience at the time. Adam was adept at making cow milk mozzarella—but knowing the challenge of working with buffalo milk, Phil sent him to Italy for the same cheese course he had taken.

Long before mastering mozzarella, Adam perfected a handful of buffalo milk cheeses for OroBianco: Buffaletta, a tangy feta-style cheese; Devil's Backbone, an aged, blue-veined cheese; Vecchio, a nutty, aged gouda-style cheese; Piccante, a sharp, manchego-style cheese; and Blanco Fresco, a rich, spreadable, chèvre-style cheese. Periodically, he makes two limited edition cheeses: Yellow Rose, an aged cheese flavored with Texas-grown saffron and black peppercorns; and a scamorza, smoked over straw from the Texas Panhandle. OroBianco's rich buffalo milk proved surprisingly versatile.

"We sold our first cheese in 2021," Phil recalls. But it wasn't until 2023 that Adam finally cracked the code to the elusive buffalo mozzarella.

There are a dozen or so steps for turning raw milk into mozzarella, including: low temperature pasteurization; adding culture for ripening and rennet for coagulating; pressing the curds; and spinning with hot water to stretch into cheese. Besides following the steps diligently, Adam learned how to account for two ever-changing variables: the volume and composition of the milk, which fluctuate with the weather. "Every batch is different. You can't just follow the recipe, you have to check things along the way," Adam says, noting that there are forty to fifty variables he can adjust when making the cheese. Because the mozzarella is made completely by hand, the entire process takes eight to nine hours, on average.

Ciao!
Y'ALL

Adam and Phil are thrilled with OroBianco's buffalo mozzarella. "I think it's on par with what they make in Italy," Phil says. "I don't say that lightly; that was after three years of me telling Adam the cheese wasn't good enough."

Buffalo mozzarella is now OroBianco's best-selling cheese. Its pleasant grassy notes on the palate—reflecting the herd's diet of mostly Coastal grass—convey its Texas terroir. The cheese also has the tender, creamy texture for which Italian buffalo mozzarella is prized. Produced every week, it outshines mozzarella shipped from Italy in a key category: freshness. "Imported buffalo mozzarella is already (at least) two weeks old by the time it hits an American store shelf," Phil says. "In Italy, they will not sell you mozzarella that's more than a day old."

These days, OroBianco's biggest cheesemaking challenge is predicting the buffalo milk volume, especially in the winter and summer. Sudden cold snaps and heat waves can reduce milk production, making it difficult to fill cheese orders, Phil says. Whereas cow milk cheese producers can buy milk from more than one dairy, OroBianco has no supplemental milk source—it's the only buffalo milk dairy in Texas.

"It's easy to make a sub-par mozzarella; it's really hard to make good mozzarella."

Phil Giglio | OroBianco Italian Creamery

Phil is proud of OroBianco's evolution as a retailer since 2021. Besides cheese and gelato, the creamery store sells butter, caramel, and charcuterie made from its water buffalo and hogs. A second location in Stonewall, Texas, sells pizza as well, topped with OroBianco cheese and cured meats. By the end of 2024, Phil expects to have three more OroBianco locations. OroBianco's cheeses—including an oil-packed buffalo mozzarella—are also sold via its online e-commerce site and at a few cheese shops and artisan food vendors in Austin, Houston, San Antonio, and the Hill Country. Restaurants in these cities—even one in Mexico—also order the cheeses. The buffalo mozzarella packaged in brine is sold only at OroBianco locations.

Between OroBianco's knowledgeable sales staff, views of the creamery through a window at the Blanco store, and photos on store walls—of the water buffalo and cowboys who care for them—Phil hopes to engage customers on more levels than taste and texture.

"We want to tell you the story of what we do, how we do it, and how it's different," he says. "We are the only people in Texas who are working with this animal, and we control everything from birthing to milking, to production, to retailing. It's a unique thing, and something our guests really appreciate."

◀ *opposite clockwise from top left*

Filatura stage (or "spinning")

Before pulling the mozzarella balls

Pistachio gelato made with water buffalo milk

Pizza mozzarella in brine

*written by* Tina Danze

*owned by*
Amelia Sweethardt and Ben Guyton

*cheese*
Chèvre, Feta, June's Joy, Silver Lining

# Pure Luck Farm & Dairy

## *Dripping Springs, Texas*

The farm was named because the owners say it was pure luck that it worked—but luck is no longer the driving force behind this award-winning dairy in Dripping Springs.

There were 1,454 entries to the 2023 American Cheese Society Competition, an annual event held each summer to celebrate excellence in cheesemaking. A small dairy farm located in Dripping Springs, Texas, tied for third place in "Best in Show" with its handmade artisanal Basket Molded Chèvre, and its Silver Lining lactic cheese won first place in the goat milk cheese category.

The creamery that produced that cheese, Pure Luck Farm & Dairy, began its life in 1979 when founder Sara Sweetser, along with two of her daughters, Amelia and Gitana, decided to purchase eleven acres of Texas Hill Country land and become farmers of cut flowers, culinary herbs, and vegetables.

◄ Amelia Sweethardt with freshly made chèvre draining in cheesecloth bags, a crucial step in the four-day cheese-making process

"My mom married my stepfather in 1983," explains daughter Amelia Sweethardt (pronounced "sweetheart"), co-owner with husband Ben Guyton of the current dairy. "She had a small herd of goats before she met him, and she always homesteaded. She was licensed as a dairy in 1995, but we continued with everything for the next decade or so. The culinary herbs that we sold dropped off about seven or eight years ago."

Sara Sweetser passed away in 2005, but not before sharing her love of creating cheese with Amelia. Two years later, Amelia married Ben Guyton, and he quickly became part of the team.

"Today, my husband does a majority of the goat milking and the herd management, and that's covering the nutrition and the housing, and making breeding decisions and plans for the projection of our herd, which is really the most important thing for the operation," Amelia says. "I am the cheesemaker, so I get to do all of the cheesemaking."

Pure Luck has about seventy Alpine and Nubian goats on the now 12.5-acre farm. The number of goats varies from year to year based on the strategy for producing milk.

"The majority of our herd are females that are lactating and producing milk, and they're on the farm for their entire lives, about ten or twelve years. There's a handful of bucks for breeding and genetics. And there's always a young up-and-coming group of milkers that are being raised as replacements for some retired does."

"Terroir" is a term most often associated in wine production, but it's also important for cheese. A French term, it is used to describe any environmental factors that affect a crop's observable characteristics.

"It's what we call 'a taste of place,' the factors that contribute to your product made on your part of the world, your piece of land," Amelia says. "What we know about goat-raising is that the Hill Country is really wonderful. It's dry and rocky, just like so much of the Mediterranean and the parts of France and Spain where wonderful goat cheeses come from. It's a good place to raise goats. It's easy to keep them healthy, and they love a lot of the vegetation. And dairy animals are like athletes or pregnant women–they need to be fed very, very well."

The chèvre made at Pure Luck is based on traditional French lactic goat cheeses. A lactic style of goat cheese is soft, tart, tangy, and creamy. Unlike hard cheese that's made in one day but won't be delicious until after aging for a few months, goat cheese production is longer, but it's ready to eat in a matter of days.

◀ *opposite*

Carrying on the Pure Luck legacy of artisanal cheesemaking are sisters Hope Boatright (left), and Amelia, with Ben Guyton and a furry friend

Ben milks the goats once each day and the milk is stored in a refrigerated tank and collected for a few days. The cheesemaking process begins with the milk being pumped into a pasteurizer where it's heated up to about 150 F degrees for thirty minutes.

"That's the pasteurization process, and it's proven to kill off pathogenic bacteria," Amelia says "Once it's cooled down, I add the culture to start the bacterial process to ferment the milk sugar into lactic acid. The second ingredient I add is a coagulant, an enzyme, to help the proteins come together to form a curd."

In the U.S., about 99 percent of all unaged, rindless cheese—like chèvre, feta, mozzarella, or ricotta—is pasteurized. Over the past decade or so, goat cheese has enjoyed a renaissance because of consumers looking for alternatives to products derived from cow milk. Goat cheese also boasts high levels of minerals and vitamins.

"On day two, we come back and it's a beautiful curd and it smells like yogurt. We drain it by scooping it out—think of a big ice cream scoop—and gently place it into perforated cheese molds," Amelia continues. "For the next 24 hours, the curd starts to drain the whey and pulls together as a little cake. On the third day, we salt it, which stops the fermentation and brings out the flavor. It's left to drain, and on the fourth day, it's packaged."

Many cheesemakers will tell you that making cheese is a three-day process but that's not what you'll hear around Pure Luck.

▼ *below*

Herd manager Ben brings the herd of Alpine and Nubian goats down from the pasture to parlor for milking

"What's unique about our cheese is that we're able to coax flavors from a very simple milk into a remarkable cheese. And that takes four days," Amelia says. "I could make this happen in a three-day process, but I like to push the limits to get the flavors I want. From a farmstead perspective, that's what I believe sets my cheese apart, making it consistent in flavor across the seasons and over the years."

Pure Luck produces about five hundred pounds of chèvre each week. From the basic white goat cheese, there are several seasoned variations including Anaheim red chili, cracked black pepper, mixed herb, and best-seller June's Joy, named for Amelia and Ben's son, made with black pepper, local wildflower honey, and thyme.

"We also sometimes make a Greek-style feta, and also a mild blue cheese called Hopelessly Blue. There's an overlapping process," Amelia says. "We don't wait to finish one batch before we start another."

Despite California and Wisconsin leading the way for cheese production in the U.S., Texas is fast becoming one of the leading goat cheesemaking states.

> "What's unique about our cheese is that we're able to coax flavors from a very simple milk into a remarkable cheese."
>
> Amelia Sweethardt | Pure Luck Farm & Dairy

"Over the years, I've watched Texas cheesemakers win awards alongside every state in this country. We've got cheeses here in Texas that are unique and represent this part of the world," Amelia says.

While small farms like Pure Luck face the high cost of being a small producer compared with higher volume creameries, Amelia and Ben meet those challenges head-on. "We sell everything we make," she says. "We're lucky we have enough people who love our cheese, and we have a great relationship with our retail partners."

It's safe to say Pure Luck's cheese is indeed very well-loved, as the creamery took home even more awards in the 2024 American Cheese Society Competition—the Basket Molded Chèvre won again, this time earning second in "Farmstead Cheeses," while its goat milk feta won first place in the goat milk category.

"Dairy is all-consuming. Recently, we've shrunk a little bit and have smaller production, because we're looking for a little more balance in life," Amelia adds. "I have to say, though, that it's quite a miracle to have a relationship with these animals—to raise them and milk them, and see our cheese become so well-received. That's one of our biggest rewards."

◄ *opposite clockwise from top left*

Salting the chèvre to enhance flavor and slow fermentation

Basket Molded Chèvre

Greek-style feta with olive oil and fresh oregano

*written by* Roger Munford

IR090623R120
IR090623R120
IR090623R120
IR091323R115
IR091323R115
IR091323R115
IR092723R120

*owned by*
Susan Rigg

*cheese*
Caldera España, Indigo Ridge, Keystone,
The Welshman, Whey Blue

# River Whey Creamery

## *Schertz, Texas*

With an entrepreneurial mind, a passion for food, and a fervent desire to learn, Susan Rigg found her way to artisan cheesemaking.

Susan Rigg is a cheesemaker and a storyteller. Behind each artisan batch of cheese crafted in her Schertz, Texas, creamery, River Whey Creamery, there is a story of Old World tradition, respect for an age-old process, and a bit of family history interwoven in every creation.

Since 2014, Susan and her team at River Whey Creamery have been bringing never-before-seen cheeses to the market and turning heads.

You might say cheesemaking is the third chapter of Susan's storied career, although those who know her best would say it was no doubt foreshadowed along the way.

After moving from California to Texas in the mid-1980s, Susan was pursuing a pre-medicine program at The University of Texas at San Antonio. The pieces, however, didn't seem to fit.

"My dad said to me one day, 'Why don't you just take a year off and work?'" she says. "I was already working part time for Barnes and Noble." Her boss jumped at the chance to move Susan to a full-time supervisory position. For more than a decade, Susan was a part of the Barnes and Noble and then Borders Books teams. At the peak of her time there, she oversaw opening new stores across the country.

◄ River Whey's Indigo Ridge ages in its specially made purple wax

Fast forward to the early 2000s, however, and Susan saw the writing on the wall amid shifts in the book industry as the online world of bookselling began to take hold. She started to look toward her next career chapter, hired a life coach, and began journaling. It always came back to her passion for food.

An endcap display for a book from the Culinary Institute of America (CIA) advertised an essay contest, and the grand prize was a week-long trip to the Hyde Park, New York, campus. As an employee, Susan wasn't eligible to enter, but her life coach suggested, "Why not just go?" So, in 2006, she left her job in the book world and headed to a culinary bootcamp in New York. While there, she learned a San Antonio campus was about to open as part of the CIA. Susan would then be a part of their inaugural graduating class, as she forged ahead to a new career, this time in the culinary world. Armed with an entrepreneurial mind and a fervent desire to learn, Susan found her way to artisan cheesemaking.

"The art of cheesemaking kept coming up for me, because it is so ancient and humans today have lost a lot of those skills," Susan says. "Thankfully, today American artisan cheesemaking is growing, especially here in Texas."

Farming wasn't in her wheelhouse, but she had access to licensed dairy farmers producing raw milk in Texas, like her current partner-producer Four E Dairy in Lavaca County. While juggling a job with Slow Food South Texas knee-deep in the regional culinary world, Susan began product developing cheese in her dining room and building out a creamery northeast of San Antonio in Schertz. Susan opened the doors of River Whey Creamery in October 2014 with three cheeses in her arsenal, all stemming from traditional ideas of cheesemaking with her unique Texas twist.

The name River Whey pays homage to the River City and its anchoring body of water, the Guadalupe. "I was driving one day to the farm to get our milk, and I drove across the Guadalupe River," Susan says. "There was a sign that said something about the 'way of the river' and that was it. I guess it was just the play on words. Instead of the way of the river, I used whey."

◄ *opposite*

Ashton Inskeep (left) and Robin Merkin cut curds in the vat Beyla, lovingly named after the Norse goddess of bees

River Whey Creamery now has five beloved cheeses, including the three they first began producing when Susan opened the doors in 2014: Keystone, Caldera España, and The Welshman. Caldera España is a Spanish-style cheese, but it's made with cow milk and smoked, differentiating it from a traditional manchego or idiazabal. "We also smoke it with pecan wood, because pecan trees are native to South Texas," she says. "What was driving me was creating something new out of an old idea.

"With each cheese," she continues, "we create something that has never existed in the marketplace before, and we tie it back to being relevant to the Texas market." A slogan of the creamery doubles down on this idea: Great Texas soil makes great Texas grass, which makes great Texas milk, and it's our job to create great Texas cheese.

The Keystone, Susan's version of an Italian tomme-style cheese, is cave-aged for four months and rubbed while aging with Texas olive oil. "When you build an arch, the piece that's in the middle, that holds it all together, is a keystone. And that was, for me, what this cheese is for River Whey. It's a simple concept as a cheese, but, boy, I'll tell you what, it is very complex in its taste, and it is a powerhouse of flavor."

Because cheesemaking is as much about paying homage to tradition and heritage, Susan developed her third cheese, The Welshman, in honor of her dad. Inspired by a cave-aged caerphilly cheese—a hard, crumbly white cheese that originated in Caerphilly, Wales—The Welshman is made using Welsh sea salt called Halen Môn, harvested on the Isle of Anglesey taken directly from the Menai Strait.

▲ *above from left*

Hoops are used to mold the curds

Wheels of Keystone float in brine

River Whey Creamery
San Antonio Express-News
JUDGING AND COMPETITION
River Whey Creamery
PAVED the WHEY
River Whey Creamery

▲
*above*

Indigo Ridge is paired with a zesty whole grain brown mustard to compliment its nutty flavor

The creamery's best-seller, Indigo Ridge, launched in November 2017, and is a cross between a gouda and gruyère. It's Susan's longest-aged cheese and comes in a specially made indigo-colored wax to complement its name. It was followed by her take on a traditional blue cheese, Whey Blue. Susan describes it as a blue cheese for a Texas palate, meaning it's not as strong a blue as you might sample in France, but it still gives you that complex profile.

Unsurprisingly, Susan's passion behind each cheese has garnered fans across the country and accolades from a myriad of notable cheese entities, like the coveted American Cheese Society.

"There is this entrepreneurial spirit, that spirit from the ones who say, 'Let's jump off the cliff and build the plane on the way down,'" she says. "That is the mentality here in Texas and where I grew up in Silicon Valley. That was also prevalent in the history of my family when they emigrated from Britain to California with very little."

For Susan, the success of River Whey Creamery goes back to her roots in the tradition and importance of food.

"I love teaching people about food and why food matters," she says. "I love bringing new products to the marketplace that didn't exist before. The creativity, the marketing, the science behind the product development, and more. This is a skill that humans have been doing for six thousand years."

Susan and the River Whey Creamery team continue to add their chapter to the cheese-making history books.

◀
*opposite*

River Whey owner Susan Rigg stands in the doorway of her cheesemaking room

*written by*
Kate Hull

*owned by*
Stuart and Connie Veldhuizen

*cheese*
Bosque Blue, Dublin Karst, Jalapeño Cheddar, Redneck Cheddar, Texas Star, Wooly Texas

# Veldhuizen Cheese

## *Dublin, Texas*

Taking as natural an approach as possible, this family-run dairy and creamery produces raw milk cheeses from its own cows and sheep.

After years of working on a dairy farm, Stuart Veldhuizen swore he would never milk another cow a day in his life. He grew up in the business and "dairied" until 1996, and his hands were tired.

But in 1999, Stuart and his wife, Connie, decided that maybe they could put their Dublin, Texas, farmland and Stuart's knowledge to use.

"My motivation for getting into cheese was not necessarily because I love cheese," Stuart says. "It was honestly a lot more so I could stay home and make the farm work, so I could help raise the family."

So Stuart's hands got to milking again. Now, more than twenty years later, he and Connie have grown Veldhuizen Cheese into an operation covering more than two hundred acres, up to 55 head of milking cows, and more than twenty types of cheese, including cheddar, blue, and havarti.

◄ (from left) Connie and Stuart Veldhuizen with daughter Rachael Gwassa

Stuart credits Connie as the "great researcher" who led them to cheese. When they were considering options for the farm, she showed Stuart producers who were making raw milk cheese on the East and West Coasts. But back then, according to Stuart, no one in Texas was producing it.

"'Gosh,' I said, 'why isn't somebody doing it here?'" Stuart says. "Well, I still don't really know why nobody was, but we started to pursue it then in January of 2000."

He began working for another local dairyman trading labor for calves, converted the Veldhuizen milk room into a cheesemaking facility, and traveled to River Falls, Wisconsin, to learn the basics, along with some more advanced courses on small farms out east.

Raw milk cheese is made with unpasteurized milk, meaning the milk has not been heated before it's used to produce cheese. Stuart and Connie noticed that those who wanted raw milk cheese really wanted it and would seek it out.

"I thought, being a small producer, I need to produce something that is, first of all, value-added, and then something that is more unique and a little more scarce, so that people will want to purchase my cheese instead of just going to a grocery store," he says.

Stuart made the first Veldhuizen cheese in the spring of 2002, a raw milk mozzarella. ("It actually worked my first batch. It was quite amazing.")

▼ *below*

The Veldhuizens' herd includes Jerseys, Guernseys, and Holstein Jersey crosses

► *opposite*

Rachael's flock of sheep is around two hundred strong

He moved on to a gruyère-style cheese, which he says there wasn't a huge appetite for, then started making cheddar due to many requests at the farmers markets where they were selling.

Over the years, the Veldhuizens have honed their craft. Milk from the grazing Veldhuizen cows is stored for a couple of days before being warmed and receiving bacteria cultures for acid production and flavor. The cheesemaker adds the coagulant rennet to help the cheese curdle, then the process changes slightly depending on which cheese is being made. Some have flavors added, like garlic or jalapeño, and some are left to stand on their own. Each cheese is aged for at least sixty days in a concrete cheese cave, though Stuart says the average aging time for their cheese is closer to eight months.

"We try to do everything as natural as possible and take care of our land, take care of our animals, take care of our family, you know?" Connie says. "Do things how I feel most people would like to see them done."

The Veldhuizens mostly sell their cheeses wholesale to high-end restaurants, hotels, specialty cheese shops, and distributors who take their products to Texas's major cities. Their Redneck Cheddar, where the curds are soaked in a Texas beer, has become one of their most popular cheeses.

And with their raw milk and cheesemaking process, Stuart says you should be able to tell the difference between a Veldhuizen cheese and a slice of sharp cheddar from a grocery store. The most discerning connoisseurs may even be able to pick out some of the Texas terroir.

"The flavors are definitely much more intense," he says. "You'll taste a lot more of what the cows are consuming. Like when there's really nice green grass, you'll taste more grassy notes and stuff like that in the milk that you won't get in a pasteurized cheese."

Veldhuizen Cheese is also very much a family operation. Two of Stuart and Connie's daughters work on the farm, as well as one of their sons, a son-in-law, and multiple grandkids. Their daughter Rachael has started her own project under the Veldhuizen Cheese name making sheep milk cheese from the more than one hundred sheep under her care, and their ten-year-old granddaughter Lexi is already an expert behind the cheese counter, Stuart says. But there are challenges being a small cheese producer in Texas—mainly the heat and drought that so often plagues the state.

▲

*above*

Dublin Karst, with its slightly fruity profile, pairs well with fresh figs and apples

"Because we're a farmstead operation, it affects our milk supply. And the cows don't tend to like the extreme heat any more than we do," Connie says. "Since we basically have a grazing operation, that changes things when we buy hay."

The Veldhuizens have been able to adapt when the going gets tough, though. In 2020, when the pandemic hit and their usual whole-wheel market dried up, they pivoted more toward retail sales and started a cheese club that is still going strong. Plus, Connie says they always have the support of their neighbors.

"On the marketing side, I think Texas is a good place. Texans like to support fellow Texans," she says. "At Christmas, we have a lot of people that'll say, 'I'm sending some to my relatives, because they've got to know we make good cheese here in Texas.'"

And even if Stuart is still getting up every day to milk cows, he's not upset about where his former life as a dairyman has taken him and his family.

"I enjoy cows," he says, laughing. "They're a beautiful animal, and they give amazing milk, and we get this incredible product. It's just fabulous."

◀

*opposite clockwise from top left*

Cheesemaker Emily Winkler-Rake drains whey

Packing Bosque Blue curds

Stuart "cores" a wheel of Huizen Kaas, a Veldhuizen signature cheese, for sampling

*written by*
Abigail Rosenthal

▲

The Four E Dairy
herd of Jerseys

6510
6367

# Even More Texas Cheese

Four more wonderful Lone Star creameries to discover.

## Brazos Valley Cheese

Waco, Texas

The cave-aged, raw milk cheese-making happens within the borders of the agrarian-centered Homestead Heritage, an intentional Christian community that stresses simplicity, sustainability, and quality craftsmanship. Ghost Pepper Cheddar, Blueberry Havarti, Horseradish Pecan Cheddar, and Truffle Gold are just a few of the incredibly popular cheeses offered here.

## Cheesemakers Inc.

Montgomery County, Texas

This family-run creamery, founded in 1997, produces and distributes cheeses on a national scale. Founder and CEO James Keliehor followed in the footsteps of his Czech ancestors, who migrated to Texas a century before. Two top-selling cheeses, named for Keliehor's twin children, are the Jaimito queso fresco and Little Miss Lonestar, a chèvre.

## Stryk Jersey Farm

Schulenberg, Texas

Founded in the 1940s, the Stryk Jersey Farm is located between Houston and San Antonio. Today, Bob and Darlene Stryk own and operate the 310-acre farm, where their Jersey cows are pasture-raised and grass-fed. Varieties of cheddar and pepper jack are made from pasteurized milk and hand cut into blocks or special shapes like the state of Texas, a longhorn, and a cowboy boot.

## Full Quiver Farms

Kemp, Texas

The multi-generational Sams family farm, located southeast of Dallas, makes artisan cheeses with milk from their own dairy herd of Jerseys and Holsteins. Their offerings include cheddar, mozzarella, colby, and pepper and monterrey jack made from raw milk. Purchase these cheeses and much more at their farm store, open every day but Sunday, and online.

## Where to Taste Texas Cheese

Visit the farmstead creameries with retail operations mentioned in this book, or check out local farmers markets, where many creameries regularly sell their cheese.

Notable cheesemongers also offer the best of Texas cheese, including:

### Antonelli's Cheese

Austin, Texas

*antonellischeese.com*

### Houston Dairymaids

Houston, Texas

*houstondairymaids.com*

### Otto's Cheese Shop

New Braunfels, Texas

*ottoscheeseshop.com*

### Scardello Artisan Cheese

Dallas, Texas

*scardellocheese.com*

## The Biggest Farmers Markets in Texas

Nearly every major Texas city offers a weekly farmers market. Find a list of some of these markets at *traveltexas.com*.

Texas Department of Agriculture offers a map of Certified Farmers Markets in Texas at *texasagriculture.gov*.

# *Perfect Pairings*

There are no rules when it comes to pairing Texas cheeses with other tasty treats. Here are some favorite combinations, suggested by Bob and Rachel Pineda of Otto's Cheese Shop. Complex or nuanced, all of these pairings are delicious!

Ash-finished chèvres like Pure Luck's Silver Lining pair well with honey.

Fresh cheeses like the Seasoned Feta from CKC Farms taste great with olives and a baguette.

Chèvres like the Goatilicious Fig & Black Pepper Chèvre compliment an arugula salad.

Soft-ripened cheeses like
Armagh's Doublin' Down
pair well with jam
and crackers.

Blue cheeses like
River Whey's Whey Blue
taste wonderful with salami.

Hard cheeses like
Haro Cheese's Eldo enhance the
flavor of a Texas Tempranillo.

# Glossary of Terms

**Aging** As cheese ages, it loses moisture while enzymes and microbes ("friendly bacteria") continue developing within the cheese. The friendly bacteria transform lactose into lactic acid and build other amino acids to give aged cheese its unmistakable taste. Most experts consider cheese to be truly aged if it is cured for three to six months. In Texas, raw milk cheese must be aged a minimum of sixty days before selling.

**Aging cave** With their cool temperatures and consistent humidity, caves have been great environments to age cheese in for centuries. Today, an aging cave can be a man-made structure with a highly controlled environment.

**Agritainment** Also known as agritourism or farm tourism, this is a broad term that refers to farm-related activities that are educational, recreational, or entertaining. The goal of agritainment is to provide visitors with a fun and interactive experience that connects them with the land, the people who work it, and the food that is grown there.

**Alpine-style cheese** Any cheese that is made using methods similar to those cheeses made in the Alps of Switzerland, including gruyère, emmenthaler, comté, and others. Characteristics include a firm yet elastic texture, low salt content, and high meltability.

**Artisan cheese** Cheese that is produced primarily by hand, in small batches, with attention paid to the tradition of the cheesemaker's art, and thus using as little mechanization as possible in the production of the cheese. Artisan, or artisanal, cheeses may be made from all types of milk and may include various flavorings.

**Ash finish** A thin layer of food-grade ash (often coming from timber or grape vines) may be used on the surface of a cheese to buffer the acidity, which creates a neutral pH environment for the mold rind to thrive, preventing other microbes from taking over the surface of the cheese.

**Cheesemaker** A person who makes cheese.

**Cheesemonger** A merchant who specializes in cheese.

**Coagulant** Refers to the addition of enzymes to milk in order to make it clot. *See Rennet.*

**Cultures** A group of specific bacteria strains that are combined in order to make a particular type of cheese. They are also used to help friendly bacteria in milk flourish and lead to a richer and more developed flavor.

**Curd** Curd is obtained by coagulating milk in a sequential process called curdling.

**Enzymes** A substance produced by a living organism, which acts as a catalyst to bring about a specific biochemical reaction. Rennet is a set of enzymes used to coagulate milk.

▲ Ash finish

▲ Curd

**Farmstead cheese** Cheese that is produced from the milk collected on the same farm where the cheese is produced.

**Fermentation** In the cheesemaking process, the baterial cultures ferment milk's lactose to produce lactic acid. The longer a cheese ages, the less lactose is left over because it has been used by the bacteria. Fermented dairy products, such as cheese, can provide numerous health benefits, including improved gut health.

**Goat terms: Buck, Doe, Kid** Male goats are called bucks or billys. Females are called does or nannys. Immature goats are called kids.

**Grass-fed** Grasses and forages (like hay and alfalfa) are the feed source consumed for the lifetime of the cow, goat, or sheep rather than grains, corn, and soy, or growth hormones and antibiotics.

**Homogenized** The process of separating fat by blasting it through a nozzel at high pressure and temperature, transforming the fat into small undetectable particles that pass straight by one's taste sensations unnoticed. This process gives milk its white color and smooth texture.

**Lactic-style cheese** Cheese made with little to no rennet, which relies primarily on the action of the bacteria converting the milk lactose to lactic acid. When the milk acidity is high enough, the milk will coagulate even without the use of rennet. Lactic-style cheese can be very fresh tasting, high-moisture cheeses with a very high lactic acid flavor.

**Microbes** An organism of microscopic size, such as a bacterium causing fermentation.

**Pasteurization** A process named after French scientist Louis Pasteur that applies heat to destroy pathogens in foods. For the dairy industry, this term means the process of heating every particle of milk or milk product to approved temperatures outlined in the Grade A Pasteurized Milk Ordinance (PMO) and held continuously at or above that temperature for at least the corresponding specified time.

**Pathogenic bacteria** Bacteria that can cause disease. Most species of bacteria are harmless and often beneficial, but others can cause infectious diseases.

**Raw milk cheese** Any cheese made from unpasteurized milk. A traditional way of making cheese. (Before the invention of pasteurization in 1822, all cheese was made with raw milk.) Current regulation (in place since 1949) regarding raw milk cheeses prohibits the sale of those raw milk cheeses that have been aged less than sixty days.

**Regenerative farming** A farming management philosophy that seeks to improve soil health. Adaptive grazing, no-till planting, no or limited use of pesticides and synthetic fertilizer are among practices that are used to restore degraded soils.

**Rennet** The most common coagulant is rennet, essentially an enzyme (made from curdled milk from the stomach of an unweaned calf) used to convert milk sugars (lactose) into lactic acid.

**Terroir** A French term used to describe the environmental factors that affect a cheese's taste and character, including soil, topography, climate, and more.

**Tomme-style cheese** Originating in the mountains of the French Alps, Italy, and Switzerland, tomme is a lower butterfat cheese often made from skimmed milk after removing the cream to make butter and full-cream cheese. It has a natural rind and offers up nutty, buttery, earthy flavors.

**Whey** The watery part of milk that remains after the formation of curds in cheesemaking . Rich in lactose, minerals, and vitamins, whey is often used as animal feed. *See Curd.*

▲ Goat terms: Buck, Doe, Kid

▲ Rennet

▲ Whey

Rachel and Bob Pineda

# Acknowledgments

We want to start with a huge thank you to all the Texas cheesemakers for all the incredible cheese that made this book a possibility. We owe our deepest gratitude to these makers, their families, and their teams for agreeing to work with us on this project. Their generosity, willingness, and time spent together gave us a deeper connection to their stories that we will carry with us forever.

We must give another big thank you to Marshall Tidrick for all his amazing photography skills. He encompassed our vision, and never has the art of cheesemaking been captured in such a beautiful light. We are so fortunate for his involvement with this project.

We will be forever grateful for our friendship and working relationship with Ramon Mejia—thank you for making us look like we know what we are doing. His passion for film is what drew us to work with him in a professional capacity to start, but his ability to capture our true selves and create an amazing final product is what keeps us coming back.

Without Amy Brinkerhoff and Melissa Lamoreaux, we wouldn't have made it this far. Thank you for being there to guide us, inspire us, and laugh with us during our regular meetings at the coffee shop to bring this project to light.

To our MVP at the shop, Bel Ross—thank you for holding down the fort and keeping our doors open for us while we set out on this mission to achieve our goals. We appreciate your dedication to the cheese industry and to Otto's Cheese Shop.

Thank you to Lindsey Schechter for all the advice, helping us out whenever we asked for favors, and being a force all on your own for Texas cheese.

To all our supporters who found it worthwhile to donate to this project, thank you for believing in Texas cheese and our vision.

We want to give hugs and thanks to Rachel's parents, Jim and Julie Earl, for being there to care for our children while we hit the Texas Cheese Trail. They have always been supportive in our career path and loving grandparents to Isabel and Claire.

And lastly, thank you to our children, Isabel and Claire. We love you so much, and because of you we are inspired to work so hard for a better future for all of us.

# Happy Trails

JALAPEÑO
Pasteurized cow's milk, cultures, vegetable rennet (mucor miehei), dci lipase (pregastric esterase), salt, non-fat milk) Jalapeño chiles and salt.
MOZZARELLA COMPANY
2944 Elm St., Dallas, TX 75226
214-741-4072
Keep Refrigerated
HARO CHEESE CO.
EST. 2021
BERTRAM TEXAS
CAMEMBERT
Double Cream, Soft-Ripened Cow's Milk Cheese
INGREDIENTS pasteurized cow's milk & cream, salt, culture & enzymes
Texas Gold Cheddar
Ingredients: Raw Milk, Cultures, Rennet, Salt.
GO TEXAN.
VELDHUIZEN Cheese
TEXAS FARMSTEAD
www.veldhuizencheese.com
3364 CR 299, Dublin, Texas 76446
254-968-3098
Made by hand. Blessed by God.
PURE LUCK
MOZZARELLA COMPANY
2944 Elm St., Dallas, TX 75226
214-741-4072
OROBIANCO ITALIAN CREAMERY
NET WT: 8OZ(227g)
MADE FROM TRUE 100% TEXAN WATER BUFFALO
BUFF MOZZ
100% BUFFALO MILK MOZZARELLA
CKC FARMS
17505 FM 32
Blanco, TX 78606
830-833-5669
Baby Caprino
"Farm Fresh Goat Cheese"
Ingredients: Grade A Pasteurized Goats Milk, Culture, Organic Vegetable Rennet, Salt, P. Candidum.
THE GREEK
GO TEXAN
WHOLE MILK YOGURT
INGREDIENTS:
Pasteurized whole milk, cultures
KEEP REFRIGERATED
Lira Rossa Artisan Cheese @ Four E Dairy
784 Co Rd 251
Moulton, TX 77975
www.lirarossa.com
NET WT 32oz
PERMIT NO 483804
LIRA ROSSA artisan cheese
RIVER WHEY CREAMERY
WHEY BLUE
HARO CHEESE CO.
EST. 2021
BERTRAM TEXAS
C'MON BERTO!
INGREDIENTS: Milk, Cultures, Enzymes, Salt
KEEP REFRIGERATED
Lira Rossa Artisan Cheese, LLC
784 Co Rd 251, Moulton, TX 77975
www.lirarossa.com
Net Wt: 6 oz
Permit No: 480320
ARTISAN · ORGANIC · DOUBLE CREAM
ARMAGH CREAMERY
DOUBLIN' DOWN
BRIE-STYLE
soft ripened cheese
NET WT. 8 OZ (126.56)
SINGLE ESTATE · MADE IN DUBLIN, TX
OROBIANCO ITALIAN CREAMERY
GO TEXAN.
BLANCO
FRESCO
FRESH WATER BUFFALO MILK CHEESE